Edward Southwick Philbrick

Defects in House-Drainage

And Their Remedies

Edward Southwick Philbrick

Defects in House-Drainage
And Their Remedies

ISBN/EAN: 9783744678377

Printed in Europe, USA, Canada, Australia, Japan

Cover: Foto ©berggeist007 / pixelio.de

More available books at **www.hansebooks.com**

DEFECTS

IN

HOUSE-DRAINAGE,

AND

THEIR REMEDIES.

By EDWARD S. PHILBRICK,

Civil Engineer.

Boston:

A. WILLIAMS & CO.

1876.

The question of House-Drainage has been for some years past forcing itself upon public attention, and yet the most extraordinary indifference prevails on the subject.

The following pages were originally written for the valuable Report of the Massachusetts State Board of Health. The subject discussed is of such vital importance to the public at large, and is receiving so general attention at the present time, that it has been thought best to present an edition in this form.

BOSTON, April, 1876.

DEFECTS IN HOUSE-DRAINAGE.

The following is written for the latitude and climate of Massachusetts.

It is not the purpose of this paper to prove the necessity of removing fecal matter promptly from our houses. This may be assumed as already proven. The intention is rather to point out how, in our community and under existing conditions, it can best be done. Neither does it seem worth while now to discuss the relative merits of the various systems by which this removal is accomplished in different places. Volumes have been written on this branch of the subject, and new schemes are constantly under trial. But the experience of all the large towns in this country, as well as that of most large European towns for the past ten or fifteen years, indicates the water-closet system for the removal of fecal matter, and the uniting with this of the refuse of kitchen sinks, for removal by water carriage through a system of drains and sewers to a destination suited to the locality, as best adapted to meet the wants of our people. It is, on the whole, likely to be attended with less difficulty in its details, and more efficiency when applied to all sorts of houses and all classes of population, than any other system of removal yet devised.* At any rate, this system has had a very

* Menzies, in his "Treatise on the Sanitary Management and Utilization of Sewage," page 8, says: "Looking at the question in all its bearings, I am forced to the conclusion that the water-closet system will supersede all others, while I believe that I shall be able to show that, agriculturally speaking, it is the best and most profitable."

Baldwin Latham, in his "Sanitary Engineering," page 328, says: "A good water-closet is the only appliance fit to be used within a house, for by it all matters are at once conveyed away, and cease to have the power of producing evil, so far as our houses are concerned. It is not so, however, with those systems which conserve fecal deposits within, or in close proximity to, our dwellings, as there is always danger in storing a dangerous article, however carefully we may tend and guard against its evil effects."

Mr. Simon, in his report as medical officer of the Privy Council and Local extensive application, and is widely popular. Until something better is devised, and has had the test of time to prove its worth, this will continue to be used. It has its weak points, however, and much remains to be done towards avoiding the dangers incident to its mismanagement, and towards

Defects to be pointed out.

perfecting its details. It behooves us, also, to seek to adapt it to the conditions existing in our community in the simplest and most efficient manner, so that they can be understood by any one who owns a house, or hires one.

DRAINS BETWEEN THE HOUSE AND SEWER OR OTHER RECEPTACLE.

Drains outside the house.

The prime object of house-drainage is the removal of the refuse with all possible speed. Every device by which any part of it is hoarded or retarded in or about the premises is

Cesspools objectionable.

to be carefully avoided. Hence, cesspools are an abomination.* Wherever sewers exist, they are worse than needless. The only excuse for any sort of cesspool near a house is the need of separating grease from kitchen-drains. Small, tight, brick tanks, or stoneware grease-pots, seem to be a necessary evil among a population who waste, or whose servants waste, so much fatty matter in their kitchen-sinks as ours. The best way to provide for this will be described later.

Essential conditions.

To secure a prompt and continuous flow, drains must be smooth inside, must be well laid, of a proper size, and have sufficient slope to render them self-cleansing. Where the last is not practicable, there should be provision for frequent flushing. They should also be as nearly impervious as pos-

Government Board, London, 1874, says : "The advantages of the water-closet system, where it can be adopted, and will be properly worked, are, as regards the extremely important object of getting the refuse continuously and completely removed, too evident to require advocacy. Those advantages, however, may fail to be realized if the system be adopted without due circumspection; and the conditions which ought to be kept in view in order to avoid any such failure are, apparently, these three: First, that the closets will universally receive an unfailing sufficiency of water properly supplied them; secondly, that the comparatively large volume of sewage which the system produces can be, in all respects, satisfactorily disposed of; and thirdly, that on all premises which the system brings into connection with the common sewers, the construction and keeping of the closets and other drainage relations will be subject to skilled direction and control."

* See Menzies' Treatise, page 20.

sible, to avoid contaminating the surrounding soil. For ^{Best mate-} house-drains, no material is so good as cast-iron, with calked lead joints. ·But glazed stoneware pipes, carefully put together with hydraulic cement, will make very good drains outside the house walls, if the soil is firm and not liable to settle. There is much of it made in this country; but it is mostly inferior in strength to the Scotch or English, which is imported at slightly higher rates.   Their connections or branches ^{Connec-} should never be at right angles, but oblique, so that T-joints or branches should never be used.
They always tend to produce an accu-
mulation of solid matter.   Y-joints or
branches can always be obtained (see
Fig. 1 and 2), and the position of the
drain can generally be adapted to their
use by taking a little pains.   When
being laid, a swab should always be
drawn through them, to wipe the surplus cement from the joint on the inside,
every new piece put into the trench
being strung on to the line or rattan which carries the swab,
and draws it along.   The writer has seen a good drain, which
would otherwise have been successful, entirely choked by
sewage accumulating against those burrs of cement inside
the joints, which should have been wiped out when laid.
Col. Waring recommends a hemp gasket at the joint, to prevent the cement from running through, but this cannot be
applied without shortening the joint to some extent, and
thereby impairing its tightness.   The lap is never very long,
at best, and cement is never so sure of stopping water as
when its surface is wiped, while fresh, on the side where the
water seeks to enter it.*

Y branches.

FIG. 1.—T-joint in Drain.

Inside joints.

FIG. 2.—Y-joint.

A frequent mistake is made in laying too large-sized pipes ^{Drains often} for drains, arising from the notion that small pipes are more ^{too large.} likely to be choked.   The fact is, that all increase of size above the requirements of capacity is an actual injury, by diminishing the scouring power of the current; so that, if laid

---

* A gasket, carefully applied, would tend to hold the ends concentric, and insure a continuity of the interior lines, but it should be applied with skill, and in limited quantity.

with a fall of two feet or more in a hundred feet of length, a four-inch pipe is better than a larger one for a house-drain used by some fifty persons, because, with this limited flow, the small one would scour better than the larger one. If rain-water is admitted from the roof-gutters, either for convenience or flushing, a larger size is perhaps needed, but six inches is ample, even then, for any ordinary house-roof. If the fall is less than two per hundred, flushing may be needed. Latham says that, in order to be self-cleansing, the house-drain should convey its contents at the rate of three feet per

Size of
drains.

second. To attain this velocity, a four-inch drain must have a fall of about one in a hundred, and a six-inch drain must have a fall of about one in a hundred and forty, even when half-full. As such drains seldom run half-full, they cannot be relied upon as self-cleansing, unless laid with nearly double the above rate of slope,—say two per hundred for four-inch drains, or one and a half per hundred for six-inch drains. For hotels and large establishments containing many receptacles for sewage and many branch drains, a six-inch pipe would be ample, unless rain-water be admitted from extensive roof-

When used
for rain-
water, this
governs the
size.

surfaces. In this case the size of the drain is governed— first, by its rate of fall, which is generally limited by local topography; and second, by the size of the roof to be drained. In our climate, a rainfall of at least one and a half inches per hour from the roof-surface should be provided for, adjusting the size of the drain to carry this rainfall. In such cases, the sewage can be practically ignored, for its volume is quite insignificant in comparison with that of the rain-water. The problem then becomes a question of hydraulics, and reference must be had to the governing elements and well-known physical laws, thence computing the required size.

Drains are not intended to carry broken crockery, old clothing, rags or shoes. Such things are often found in them, it is true; but increasing the size of the drain is no remedy for such abuse, which would choke a street sewer. On the other hand, the smaller the drain which will carry the largest flow with which it is likely to be taxed, the better is the scour, and the more likely it is to keep clean. Any accumulation of sewage in the pipes is sure to decompose and give rise to abundance of poisonous gas, which it is next to impossible to keep out of our houses. It cannot be expected that Sewer-gases to be shut out and diminished in volume. the interior of sewers and drains should always be free from such gases, but it is by all means desirable to reduce their volume to a minimum, and then to apply all possible precautions to prevent their mixing with the air we breathe. To prevent Traps. their access to our houses, traps are used. To a certain extent, and in certain places, they are essential, but there may be too many traps. Every trap in the line of a waste or soil pipe is necessarily a place for sewage to be arrested temporarily, and, if the use of the pipe be not very frequent, decomposition occurs, evolving gases.

In all houses draining into sewers, the place where a trap is most essential is outside of the house walls, on the main house-drain, after it has collected all the branches which are tributary to it, and between this point and the sewer. Prof. Edmund Parkes, in his treatise on practical hygiene, says, page 343 : "It is hardly possible to insist too much on the importance of this rule of disconnection between the house-pipes and outside drains. Late events [supposed to be the illness of the Prince of Wales] have shown what a risk the richer classes of this country now run, who not only bring the sewers into their houses, but multiply water-closets, and even put them close to bedrooms. The simple plan of disconnection, if· properly done, would insure them against the otherwise certain danger of sewer air entering the house. Houses which have for years been a nuisance from persistent smells, have been purified and become healthy by this means." The medical officer of the Privy Council, London, says : "This condition ought to be insisted on : that every private drain be properly trapped and ventilated in relation to the common sewers," etc. (Report of 1874, p. 32.)

In England and other places having a mild climate, it is Traps between house and sewers. usual to disconnect the house-drains from the street sewers by providing that the former should discharge their contents into a chamber or tank, open at the top, just outside the house walls, into which the rain-water spouts are often turned. The rigor of our New England winter prevents our people from following many of the devices which in Old England are quite efficient, and this one among others. All out-of-door drains are here of necessity kept deep in the ground, with as little exposure as possible to a temperature of 40° Fahrenheit below freezing, which sometimes prevails for several successive

FIG. 3.—Trap for Drain.

days.* The best sort of disconnection we can apply, is to introduce a pipe-trap between the house and the sewer. This should not be a built chamber with square corners, which might collect solid matter, but a mere depression in the pipe itself, having the same sectional area as the pipe, and therefore containing the minimum of matter for decomposition. (See fig. 3.)

Such traps may sometimes be forced by the compression of air in the street sewers, especially if these are tide-locked

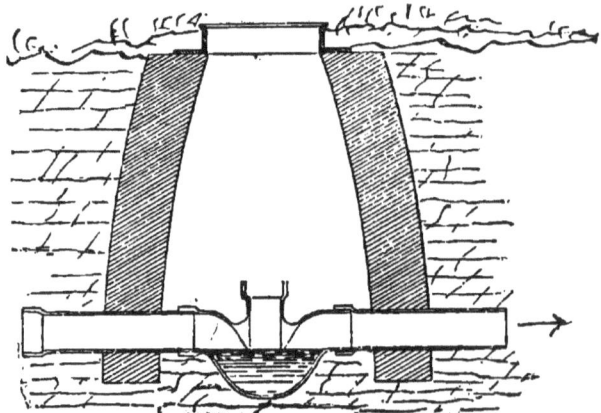

FIG. 4.

Vents for outside traps.

at high water, like many of those in Boston. To provide against this, a vent-pipe, of four inches diameter at least, should in cities be led from the hole in the trap directly up the side of the house, like a water conductor, and above all dormer windows. The water conductor itself will not answer for this purpose, for the compression of air in the sewer is most likely to occur during a heavy rain, when the water-spouts are fully occupied as such, and are, therefore, incapable of giving vent to the gas, for which special outlet must

* At the moment of writing this, November 30, 1875, the thermometer has averaged only two degrees above zero, Fahrenheit, for the past twenty-four hours, with a gale of wind from the north-west, forcing the air into every crack and cranny of our houses.

be given. In suburban districts, a vent into a pile of loose stones, or a man-hole chamber under ground will answer. (See fig. 4.) During the winter this chamber may be filled with dry leaves, etc., and the vent covered with wire netting, in order to prevent freezing.

A perfectly ventilated system of sewers would doubtless render this vent-pipe needless ; but few of our towns, if any, have attained this stage of perfection in this respect. The method of sewer-ventilation advocated by Baldwin Latham, and largely practised in England, and by J. H. Shedd in this country, consists of small holes in the man-hole covers in the streets. In our climate, such vents are completely sealed by ice or frozen mud for six months, and perhaps by liquid mud and dust for a large part of the other six, unless cared for by men kept for the purpose. Sewer-ventilation generally imperfect.

The subject of cesspools has been alluded to above. Even where no street sewers exist, the cesspool may sometimes be dispensed with. Col. George E. Waring, Jr., in a series of excellent articles lately published in the "Atlantic Monthly," tells of his own experience in distributing his sewage through the soil of his lawn by porous pipes, serving to utilize the material in the simplest and cheapest manner. The writer has pursued a similar plan for over twenty years successfully. But many people have too little land about their houses to provide even this small "sewage farm" within their own limits.

The cesspool, then, in the absence of sewers, becomes a necessity, and large numbers of our people are thus driven, by the increase of poulation, to live on quarter-acre lots, and even smaller ones, with their old privy-vaults, cesspools and wells for drinking-water within one or two rods of one another ! The habits of our people demand all "modern conveniences" inside their houses. They ask for water-supply and waste-pipes in all directions, and upon every floor ; but if dependent upon wells for their drinking-water, these sources are sure to become sinks in the course of time. Neither can they expect any warning, appreciable by the senses. The change is insensible and invisible. The well-water may look as pure as ever, and taste as cool and refreshing, and yet contain the seeds of disease. Some argue Poisoning of wells.

that because their cesspool is on lower ground than the well, the latter cannot be affected, for, say they, "the dirty water can't run up hill." They forget that the contents of the cesspool may be twenty or thirty feet higher than the bottom of their well, from which they generally get their supply, and that although underground drainage, which supplies the well, generally runs in a direction indicated by the slope of the surface, there is no certainty about its always being so. There are times when the well-springs are low, and but little water is found in them. How do they know, then, that the cesspool, though in ground lower than the top of the well, may not soak in the direction of the well, whose bottom, nearly empty, is many feet below it? If the soil be once polluted about the house below the absorbing powers of surface vegetation, whose roots seldom go deeper than one or two feet, it never can be relied upon again with safety for the filtration of water for drinking.

The accumulation of filth in the soil around these porous cesspools is just as certain as the annual rise of the streams after the winter rains, and such accumulation is as certain to be followed by injurious effects upon the health of people whose houses are near such influences, as is any other violation of sanitary laws. It is said to be the invention of the shirt that brought us immunity from the plague, through the improved cleanliness of the skin. But if our people go on as they have done to pollute the soil about their houses by using water, as they now do, to dissolve their filth, rinse it out of their houses, and soak it down into their soil, the most frequent changes of linen will not save them. They must invent some other source for their drinking-water, than to pump it up again from the same soil, or the plagues of the East will visit us again in some form or other.

Sewerage imperative with an abundant water-supply.

SEWERAGE should follow immediately, or be provided simultaneously with water-supply. For, if wells are abandoned, and aqueduct water is to be substituted, the consumption of water is multiplied at once, and cesspools become quite inadequate to dispose of the house-washings. If sewerage is not provided simultaneously with a water-supply, the surroundings of our houses soon become saturated with water as well as filth, and steam up, under our July suns, to infect our systems, through the lungs instead of the stomach, with consequences quite as fatal, and probably more speedy.

The details of the construction of sewers, and the ultimate disposition of the sewage, are subjects demanding a separate study for each new locality, and their investigation would be beyond the limits of this paper. No branch of civil engineering is more important, or more fraught with difficulties demanding skill and a careful study of the experience of others. It is, in fact, one of the most important questions connected with the growth of our modern civilization. No community can afford to ignore it.

Trouble often arises from the settling and breaking of house-drains, when laid upon filled land. The books say that they must be laid in "virgin soil." It might puzzle the wisest to find any soil to answer that description among the many thousand houses built upon pile foundations in Boston and its suburbs. The occupants of such houses at the South End of the city, and in the fine mansions on the "Back Bay" have had a good deal of trouble from this source. In fact, it must exist, in some degree, over the greater part of wards 7, 8, 9, 10, and 11 (old divisions). The houses being built and occupied long before the mud bottom under the filled streets has become thoroughly settled, this process of settling continues in some places for years, carrying down with it the house-drains, which are inevitably sheared off near the outside of the house walls, for these are built on rigid foundations. The immediate consequence is a leak in the drain close to the outside of the cellar wall, and in some cases, entire breach of continuity. *Drains laid in filled land.*

If leaking alone, the only warning received by the occupant of the house is in the percolation of the sewage through the wall or up through the cellar floor, for there is not one wall in five hundred that will stop it. Neither will concreting cellar bottoms stop it. The more resistance there is offered to the influx by such walls and floors, the more the filth is accumulated in the surrounding soil by lapse of time and constant leakage from the cracked drain, till the clean, porous gravel with which the street was once filled becomes saturated with the sewage, a sponge of an uncertain extent, filled with the foulest of matter, which it is next to impossible *Broken drains.*

to shut out of the cellars, for it is both fluid and gaseous, and penetrates the minutest pores.

Remedies.   The remedy for this nuisance is by no means simple. Wooden boxes are slightly pliable, and, if made with care, and well clamped, may answer sometimes for temporary house-drains, till the material under the street has ceased settling; but even wooden boxes cannot be bent far without opening joints and becoming leaky. If houses must be built and occupied in such places, the only sure way of construct- ing a permanently tight house-drain would be to drive a row of piles for its foundation, between the house and the sewer. Even then, if the sewer is not built upon piles,—and they rarely are,—the break would occur where the piles cease, for nothing else is rigid over the compressible mud of these regions. This evil is so widely prevalent, that great com- plaint has arisen about the drainage of those districts, the source of which is more likely to be traceable to cracked house-drains, than to any defects in the sewers themselves. It is certainly a serious matter for any one who contemplates living upon newly-filled lands. The use of cast-iron drain- pipe all the way to the sewer, with calked lead joints, is recommended by some authorities, in soils subject to settling. But even iron pipes will break, if rigidly connected, about as soon as stoneware, though, having fewer joints, they may break in fewer places. They are certainly no sure remedy for this evil. If a tight, flexible pipe could be made, it might answer the purpose for awhile, but such a thing is yet to be invented in a permanent form.*

Man-holes for access recommend- ed.   A commission recently appointed by the city government of Boston to consider the drainage of that city, recommend making a man-hole for access to the house-drain close to the outside of the house wall, so as to allow of ready inspection for detection and mending of leaks, caused by settlement of newly-filled lands. This is an excellent suggestion, and if the leaks were confined to this point, would help the case materially. This point is the one where settlement is most likely to occur, and it may cover the whole trouble in a

---

* A flexible drain-pipe, made by coupling short joints of iron with rubber gaskets, if carefully put together, might answer for a number of years, but any packing of such organic matter is subject to decay, and then leakage occurs.

majority of cases, *if* well watched. Of course, it would need
protection from frost in exposed situations. This could
readily be given by filling the man-hole chamber with straw
or litter.

Where a " virgin soil " exists, there is, of course, no excuse
for the breaking of drains. Yet they sometimes do break,
from the want of care in the laying or in packing the earth
around or under them, especially where passing across the
earth newly filled around the outside of a cellar wall. Such
places should always be puddled with water when filling, both
under and over the drains. Of course, every leak is a source
of great risk, contaminating the soil in its vicinity to an extent dependent on its permeability. In short, no workman- Good workmanship is
ship can be too good to be employed in laying house-drains. essential.
They are out of sight, and, therefore, out of mind. Moreover, a defect can only be detected after months, if not years,
during which time the soil may have become polluted to an
incurable extent, rendering a home a mere pest-house which
might otherwise have been healthy.

The increased use of water in our houses is justly regarded Increased
use of water
as one of the most valuable agents in raising the standard of brings greater risks.
cleanliness among the poor, and in contributing to the com-
fort and luxury of the more wealthy. But it must not be
forgotten that it brings with it these increased risks, and
demands the most careful attention ; for the more water we
dilute our sewage with, the further will it penetrate through
pores and diffuse itself through the soil, unless securely led
off in proper channels to proper places.

DRAINS WITHIN THE HOUSE WALLS.

The above remarks apply chiefly to the drains outside of Drains
inside the
houses. But that portion of the drain which is *within* the house.
walls deserves still more rigid scrutiny. The soil outside has
certain absorbent powers, combining chemically with the poisonous gases, or holding air in its pores for their partial oxi-
dation. Moreover, the poisonous influences within the walls,
are much more likely to be absorbed by and act upon our
systems through the lungs, than those which are partially shut
out by the walls, or partially diluted by the open air. A
New England climate does not admit of much " fresh air "

The poor cannot afford fresh air in winter.

inside the homes of those who cannot afford to heat it during six months of the year. The suffering from frost is immediate, leading the poor man to calk every crack, while bad air is a slow poison, warning us perhaps by the sense of smell, in some degree, yet not in the urgent manner which would lead to an appreciation of its importance. If not immediately attended to and changed, the bad air soon ceases to attract our attention through the sense of smell, and is never thought of as a serious matter by a large part of our population. In fact, they might perish with the frost if they failed to shut out the pure air, and so choose the chance of living by shutting out both frost and air together. We must therefore expect to find poorly ventilated houses among the poor in winter. The exhalations from the skin and lungs are, unfortunately, not so easily collected and got rid of as the fluid and solid excretions of the body. But in getting rid of the latter, if we do not take great care, they, too, become gaseous, and return to plague us in the air, already heavy with the vapors from the lungs and skin in badly ventilated houses. The introduction of water-closets and slop-sinks into tenement houses should therefore be guarded with peculiar attention, or the benefits to be derived from their use will be more than cancelled by the evils which may arise from their defective construction.

"Modern improvements" in cheap houses.

A great number of houses have been built within a few years upon speculation in the vicinity of Boston and other large towns by a class of professional builders who erect long blocks with borrowed money, reducing the cost to a minimum by doing the work in a wholesale way, building by the dozen as it were. Every part of the work is subjected to competition and the lowest bids taken, regardless of the reputation of the builder. The drainage and plumbing of such houses is generally calculated to please the eye by a display of marble slabs and plated mountings in convenient places; but there being no reward offered for good workmanship or good planning, neither is to be expected. It is here we find a combination of bad designs, defective work, and poor materials, with a display to catch the eye, making a sort of man-trap or whited sepulchre; for no sooner does a family attempt to use such a house as a home, and to turn its drainage into the receptacles conveniently provided for the purpose, than we find sewer-gas diffused everywhere. The occupants of such a house would be safer, in many cases, if all their sewage were thrown into the middle of the street, or even on the sidewalks, to decompose in the sunshine, or to be eaten by the dogs and rats in true Oriental style; for the products of its decomposition would then at least be diluted and scattered by the winds, and would not be carried about their houses in concentrated form by pipes and passages, to poison the air of the bed-chamber and nursery.

Menzies says (page 13): "The gas which arises in foul drains is of a singularly light character, and has a tendency to ascend or draw towards any heated part of a house. Hence it often arises that houses in towns situated on the highest ground are more unhealthy than those in the valleys,— the foul air rises to them through the drains! As during the greater part of the year the internal temperature of an inhabited dwelling, and especially of some parts of it, is much higher than the surrounding atmosphere, it is obvious that the gas naturally ascends to the living-rooms, especially if during the winter and autumn they are warm and comfortable. These water-closets are also generally on the bedroom floor, and it is more injurious to health to sleep in foul air than to be in it during the day-time." *Lightness of sewer-gases.*

In planning house-drains, they should be got outside the walls of the house as directly as possible. In public institutions, or other large buildings, where a large number of receptacles of sewage is provided, the main drain for the collection of the whole should be outside the walls, wherever practicable, for the reason that fewer joints of pipe, and fewer chances of leakage from imperfect work, would thus occur within the walls. *Drains to be led outside the walls.*

The material for drains within the walls should be metal in all cases. It is often customary to lead a drain across under a basement floor by stoneware pipes, which, though much better than the old-fashioned brick drain, is far inferior to iron. The writer has seen such a drain, well laid with Scotch pipe and full cement joints, and covered with concrete of hydraulic cement on the cellar floor, giving off through this cement an amount of stench that made the cellar nauseous, *Material for drains inside of houses. Cement is pervious to gas.*

even though the soil-pipe above was ventilated. The sewer in the street may have been in fault, but this case serves to show how penetrating are these gases, and that good hydraulic cement mortar, though impervious to water, is not impervious to them. A ventilated trap outside the house afterwards stopped this nuisance in the case referred to, but even this may not be enough in all cases, for a certain amount of slime inevitably collects upon the insides of house-drains themselves, which, by its decomposition, evolves gases requiring metal

Sewer-gas very penetrating. joints to hold them. Menzies says (p. 14), "I have known this gas pass through floors and through chinks in two-feet walls. It will find out the smallest opening in any pipe that will give it a chance of getting to the heat or the open air." This same gas, if escaping from a slight leak in a drain buried in the soil outside the house, would doubtless be absorbed and rendered innocuous by the soil and by the air within its pores, but under a house the case is widely differ-

Iron pipes. ent. Cast-iron pipes, with leaded joints, well calked, and painted, are safe ; and unless subjected to such great changes of temperature as might loosen the joints by expansion and contraction of length, will prove satisfactory for a long term

Should be above floors in base-ments. of years.* If iron is used inside the walls, there is seldom anything to be gained by burying it under the cellar or base-ment floor. Such pipes should be readily accessible for in-spection. If a little attention be devoted to the subject, they can generally be placed along some wall or partition, or hung from the ceiling, where their joints can all be readily seen to be recalked and painted whenever necessary. If a water-closet be placed in the basement, it should be near the wall, where the soil-pipe leaves the house, so that this pipe, pass-ing just above the floor, can serve for its drainage. If neces-sary to this end, the floor of the closet can be raised one or two steps above the rest of the basement floor. Prof.

---

* Latham, in his Sanitary Engineering (p. 319), gives preference to lead over iron for soil-pipes. But the superiority of cast-iron over lead has been amply proven in this country for over ten years. Latham's objection to the rusting of iron pipes may be applicable to wrought-iron, but does not seem valid as to cast pipes, the insides of which soon become coated with a film, beyond which the rust does not penetrate far. Lead soil-pipes, on the other hand, are very difficult to secure against sagging out of place, against rats, and against corrosion, or nails carelessly driven.

Parkes says (Practical Hygiene, p. 343): "It should be a strict rule, that no drain-pipe of any kind should pass under a house. If there must be a pipe passing from front to back, or the reverse, it is much better to take it above the basement floor than underneath, and to have it exposed throughout its course."

It has been a common practice in England to provide a *Cellar and basement drains.* drain from the cellar or basement into which the scouring water can be emptied which is used in the washing of floors, etc. This drain is generally discharged into the main drain outside of the house. A trap is generally provided under the gulley or sink where the water is poured, but, as the place is not in daily use, this trap is likely to become dry or filled with sand from the scouring water, and in either case useless. It is better to dispense with the trap in such places, and depend upon the trap which should always exist outside the house. The only risk of bad gas from such a sink, would be from the other house-drains themselves. This risk is not to be ignored, and in order to escape it, such a sink should be placed, not under the house, but in an outer shed, or, better yet, in the yard, outside the house, where a grating can cover it. A pocket or catch-basin for sand should always be provided under the grating.

The drainage of the soil on which a house is built, if it *Foundation and subsoil drains.* consist of porous sand or gravel, will not require much attention, unless the level of the cellar be decidedly below that of most of the surrounding land, as in broad plains or valley-bottoms. When such cases occur, or when the soil is impervious, a porous tile-drain should be laid, three or four feet deep if practicable, with porous material over it around the bottom of the foundation-wall, with a delivery to the house-drain above its outside trap which disconnects it from the sewer. In case no sewers are provided, among a scattered population, such a drain can generally be led to some point low enough to discharge it on the surface of the same lot; if not, the lot is very ineligible for building purposes.

Branch drains from sinks, wash-trays and wash-bowls are *Branch waste-pipes.* generally made of lead, which seems to be the most suitable material. Its pliability and durability are valuable qualities. The first may lead to its distortion of form, by sagging, if not well supported. Where these lead waste-pipes enter the iron ones, a common practice among plumbers is to secure the joint by glazier's putty. This is but little better than a rag packing for such a place, for the slightest expansion and contraction of the pipes, endwise, by changes of temperature, will crack the putty and lead to its crumbling away in one year. The only proper way to make such a joint, is to solder a tinned iron or brass ferule to the outside of the lead pipe, which is to enter the bell of the iron pipe. This ferule gives a stiff material against which a lead joint can be calked in the same way as between two pieces of iron pipe. This lead packing will yield to the expansion, without breaking or crumbling. When lead traps are used under water-closets, the joint between them and the iron soil-pipe should be secured in the same way.

Putty joints.

The connection of waste-pipes from wash-bowls, bath-tubs, wash-trays, and of tank overflows, with the soil-pipes, has given rise to much trouble. In this neighborhood, it is customary for plumbers to enter them into the trap of the nearest water-closet below the water-line. This is often carelessly done, making the connection so near the surface of the water in the trap that the seal is not reliable. Moreover, the emanations of foul gas from the water in the trap would rise through the cistern overflows and render the water in such cisterns unsafe for drinking, for the rarity of the use of such overflows renders traps in them liable to dry up, and therefore of little value. The English discharge such waste-pipes over a grating in the open air, which drains into the main drain below, thus insuring complete disconnection. But this is not practicable in our climate, and we must seek some other method. The overflows of tanks must needs have open mouths at their upper ends, and should, therefore, not connect with a foul pipe below. The rain-water conductor from the roof is a fit place to discharge them, unless itself connected with the drains below. When thus disqualified, the overflows can terminate in the open air outside the house, for they will never carry water enough to cause annoyance, the water being always clean. The wastes can be safely discharged, as above described, into a water-closet or slop-sink trap, if pains be taken to enter them at its bottom, as far as possible below the

Delivery of waste-pipes from bowls, etc.

Overflows.

English method impracticable here.

water-line. If such wastes are of considerable length, say ten feet or more, they may become offensive, from the decomposition of the slime inside them in warm weather, and should have S-traps near their upper ends, with vents from the top of these traps at least one inch in diameter, connecting with the main vent of the soil-pipe. The practice of drawing drinking-water from a tank or cistern must be condemned under any circumstances. There is no reason for it, with a constant supply in the mains, such as is universally given in our country. The practice originated with the system of intermittent supply, formerly prevalent in England, but now going into disuse. *Drinking-water should not be drawn from tanks.*

Soil-pipes from water-closets were formerly made of lead, at first by soldering sheet-lead into cylinder form, and afterwards by the seamless process. The first show more rapid corrosion at the solder-joint; both are subject to corrosion and sagging, and to being gnawed by rats. Iron is much safer, and fortunately cheaper also, and is therefore now generally used. *Material for soil-pipes.*

Plumbers sometimes connect branching soil-pipes by T-joints, when it serves their convenience. Y-joints should always be used, for the same reasons as given above for connecting outside drains. The Y-joint sometimes requires the introduction of another small bend to complete the necessary change of direction. Hence arises the temptation to use the T-joints in contract-work, to save the cost of the bend and its application. *Y-joints, and no T's.*

Rain-water cisterns are sometimes built in basements, or outside of houses, underground, having their overflows in the house-drain. Such an arrangement is never safe. However carefully the overflow may be trapped, the long drouths of our climate may dry up the water-seal, and allow the sewer-gas to spread over the water and be dissolved by it. Moreover, the drain may be obstructed below the junction of the overflow, and the whole house-sewage is then backed up through the overflow into the cistern. Such an instance actually occurred within the knowledge of the writer, where about a barrel of grease was allowed to collect in the drain from the kitchen-sink, after filling the cesspool provided for the purpose. The first intimation received of this obstruction by the *Rain-water cistern overflows.*

c

occupants of the house was that the cistern-water, which was used through a filtering-pump for drinking, got a "coppery taste," as they expressed it. On cleaning the cistern, it was found that the whole drainage of the house had been emptying into it, apparently for some weeks. Cistern-overflows can generally be discharged on to the surface.

Receptacles
for grease. The need of pots, tanks or other receptacles for the collection of kitchen-sink grease, has been alluded to above, and will be generally felt in all houses where the inclination of the drain is not very rapid. The writer has tried various devices for this purpose. The subject is ever fruitful of annoyance, especially among small families who employ no man-servant, and whose members dislike to meddle with dirty messes. Sufficient space must be given for the accumulation of grease during the intervals between the times of cleaning. The inconvenience of frequent cleaning in our winters, when the congelation of the grease is most rapid, and the inconvenient depth below the surface required to escape frost, have led to putting them in cellars, and even above the kitchen floor, under the sink, in some cases. The latter was found unendurable, from the stench arising when cleaned, and the frequent cleaning rendered necessary by the limited space for storage. Stoneware pots were tried, of about ten gallons capacity, both in kitchen and cellar. When in the cellar, the cleaning is less offensive ; but in severe weather it is difficult to so ventilate a cellar as to keep out of the house all effluvia so arising, while their limited size requires attention once or twice every month in winter. In the days when our grandmothers presided over their houses, there was more attention given to the small economies of the household, either by those grandmothers in person, or by some servants who felt more impressed with the sin of wastefulness than do the denizens of our modern kitchens. Certainly there is a value in the grease now thrown away in dish-water which ought to lead to its being collected before going into the water, instead of encumbering our house-drains with it to such an extent as is now done. Until such an end be attained, however, some means must be provided for keeping it out of the pipes, for no matter how large these may be, if as large as a flour barrel, they would be filled solid with grease in one winter by some sinks, even where the family consisted of but five persons, all told. The best plan yet devised for this purpose is, perhaps, a small brick tank, laid in hydraulic cement, and plastered smooth inside, placed as close as possible to the cellar wall on the outside, with the sink as close as possible to

SECTION

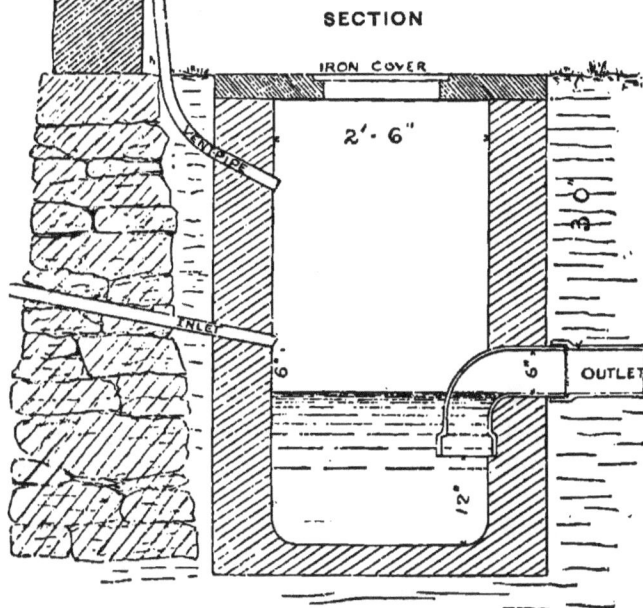

FIG. 5.—Cesspool or tank for grease.

the same wall on the inside, so that the grease shall not congeal in the pipe between the two. (See Fig. 5.) For small and medium houses it should be from one and a half to two feet square on the inside, with the bottom about two feet below the outlet-pipe, which is to turn down about a foot on the inside, with a smooth, round turn, so that its mouth may be so much under water. The inlet should be about six inches higher than the outlet-pipe, to allow the grease to collect to that

PLAN

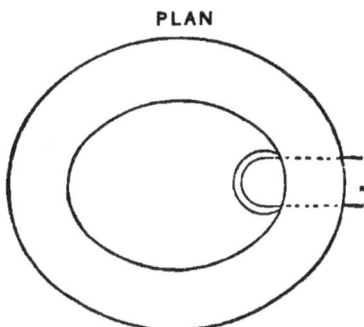

thickness above the water line, which is governed by the level of the outlet, without obstructing the mouth of the inlet. The grease will then float on the water, and become congealed in the form of a dirty scum, while the water and other matter in suspension flow out by the mouth of the outlet, about a foot below the surface. The whole must be so placed as not to freeze. The depth needed for this will depend largely upon the exposure. The walls being built to the surface of the ground can be covered with a flag-stone, with hole and iron cover.

The soil-pipes from the water-closets should by no .means enter this receptacle. It should be upon a branch drain, serving the kitchen and scullery sinks alone, having its outlet into the principal drain.* If more than one sink delivers into it, the tank itself should have a vent-pipe, to prevent the air, compressed by the influx of water from one sink, from being forced up through the trap of the other inlet-pipe into the house. If the waste-pipe becomes choked with grease between the sink and cesspool, as will often happen when the fall is not rapid, it may be sometimes kept clear by flushing occasionally with boiling water, provided the passage be not wholly obstructed.

Fixtures inside the house. Waste-pipes found in too many places.
The demand for modern conveniences has introduced waste-pipes all over our houses. Their orifices are found in bed-rooms, and on every floor, from attic to cellar. With perfection in planning, workmanship and management, such things may possibly be made safe. But we cannot expect perfection in either of these departments. The nearest approach to it ought to be looked for among the homes of the wealthy, whose means can command the services of good architects and good workmen, and who are not stinted as to the cost of

* The reason for excluding the soil-pipes of water-closets from the grease-pot or tank,—though their introduction is apparently recommended by Col. Waring, in speaking of Field's flushing-tank,—is the same as has been given against the storage of all fecal matter near a house, for however limited a period, in hot weather. Moreover, the separation of the grease is much more difficult if fecal matter be introduced into the same receptacle. Field's flushing-tank is doubtless an excellent arrangement for mild climates; but our winters would require its inlet to be placed so far below the surface, that the additional depth required to work the siphon would render its outlet inconveniently deep, except where the ground slopes rapidly, and then the flushing is of little importance, comparatively.

such appliances as conduce to safety in drainage. But it is precisely in the most costly houses that the waste-pipes are found most widely scattered. The chances of imperfect work are too great to justify the practice of putting wash-basins or water-closets in sleeping-rooms under any circumstances, or even in dressing-rooms, closets or passages leading directly from sleeping rooms, unless these conveniences are supplied with the most ample and thorough ventilation, put together in the most careful manner, and most scrupulously taken care of. English authorities say that water-closets should be built over one another in a tower projecting from the house.* The placing of one over the other is a great advantage in point of economy of construction. But the placing them outside of the line of the house-walls is hardly practicable in our climate. Our winters compel their construction, either on the southerly side of our houses or in their interior, unless special heating arrangements are applied. It therefore becomes all the more necessary to provide special means for their ventilation.

Waste-pipes not to be in sleeping-rooms.

Placing of water-closets.

The water-closet is used by thousands who know little or nothing of its mechanism, and who necessarily consider it as an automatic arrangement, needing little or no attention, and who therefore bestow none upon it. But, as it is no more perfect in its way than all other work of human hands, it has many faults and weak points, particularly in the form of the pan-closet, now so generally used. It therefore behooves the architect who plans a house for the rich man, the mechanic

General ignorance as to their mechanism.

---

* See Eassie's "Sanitary Arrangements," page 62; also the supplementary report for 1874 of the medical officer of the Privy Council (England), from which the following extract is taken, page 33 :—

"In considering the admissibility of water-closets, it has always to be remembered that the working of an ordinary water-closet is easily deranged, and that water-closets, when out of order, and especially if in the interior of houses, are apt to become very dangerous nuisances. The ordinary water-closet is, therefore, a thoroughly ineligible form of privy for those who are unlikely to take proper care of it, or are from poverty unable to give it such occasional repairs as it may require."

"Among such classes of population it is, of course, unfit that any form of indoor privy should ever be sanctioned; but even in the best-ordered houses the occasional danger of indoor water-closets must not be disregarded. Water-closets ought never to stand where they cannot have outside windows: they ought, if possible, to stand as projections from the body of the house, and with windowed lobbies dividing them from it."

who plans his own, or who builds to sell again, and lastly the householder and head of family himself, to know something of the general principles of its construction, and to avail himself of such knowledge in planning, building and taking care of a house. There seems to be a deplorable lack in this respect, for instead of closets and drains, placed so as to insure the getting rid of the refuse with safety, we often find poisonous gases emitted from them, and conducted all over the house, by an ingenious system of pipes, floor-spaces and partition-spaces in our plastered buildings.

Water-closets of ancient origin. The use of water-closets dates from a very remote period. Thomas Ewbank, in his historical treatise, says (p. 561) : "They are an ancient and probably an Asiatic device. The summer chamber of Eglon, king of Moab (Judges iii. : 20–25), is supposed to have been one. They were introduced into Rome during the Republic. Those constructed in the palace of the Cæsars were adorned with marbles, arabesques and mosaics. At the back of one still extant, there is a cistern, the water of which is distributed by cocks to different seats." Their general use in private houses dates, however, from a very recent period. It Ordinary style very defective. is to be regretted that among the hundreds of patented inventions, recently brought before the public, one of the most defective and dangerous of them all should have got into such general use in this country ; viz., the ordinary "pan-closet." Baldwin Latham speaks of them in his "Sanitary Engineering" (page 329), as "cumbrous appliances, which cannot be introduced into a house without creating a nuisance." The fact remains, however, that thousands of our fellow-citizens have already fitted their dwellings with them, at a considerable cost, and it becomes important to remedy their defects as far as possible where already in use. The following devices are recommended for those who have already made this bad investment, while advising those who are building anew to adopt some of the simpler and safer inventions which will be described afterwards.

The pan-closet is described in the annexed cut, figure 6, and consists of several parts : *First,* The bowl of crockery, directly under the seat. *Second,* The copper pan, which, when ready for use, is in the position shown by the dotted lines, is full of water, and seals the bottom of the bowl.

Third, The cast-iron receiver, standing on the floor, within which the pan is tilted when discharging its contents; and *Fourth*, The lead trap just below the floor. Its defects are numerous, but its chief defect arises from the reservoir of foul air always present in the iron receiver below the crockery bowl. The inside of this receiver is necessarily foul. It is quickly smeared with filth when first put in use; its in-

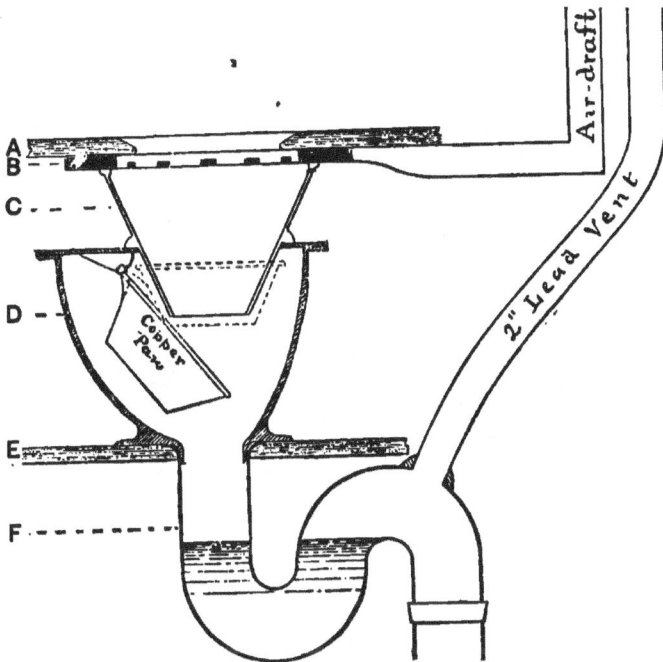

FIG. 6.—A, Wood seat. B, Annular ventilating tube. C, Crockery bowl. D, Iron receiver. E, Floor. F, Lead trap.

terior is inaccessible, and can therefore never be cleansed. Directly below it is the large metal trap, whose contents generally emit noxious gas from their decomposition, and this trap cannot be safely dispensed with. Whenever the pan is tilted and discharged, there is suddenly dropped into this receiver several quarts, and sometimes a pailful of water. This must of course displace its own volume of the foul air pent up there, for which there is no escape in any direction but upward, with a rush, past the tilted pan into the bowl, where it mixes freely with the air of the room. Various

schemes have been devised for getting rid of this nuisance.
When not gotten rid of, the pan-closet is a dangerous neigh-
bor. Several devices are described by Eassie ("Sanitary
Arrangements," p. 74) for injecting a disinfecting fluid by an
automatic apparatus into the pan or bowl at the instant of
Disinfecting opening the valve. Such contrivances may serve a good pur-
water-
closets. pose, if well regulated, but the adjustment of these additional
parts, and the occasional renewal of the disinfecting agent,
complicates matters somewhat, and renders such a remedy
less simple and less adapted to general use than it is desirable that it should be.

Ventilating If the water-closet can be located near a chimney, which is
water-
closets. sure to be in constant use, as the kitchen chimney, the evil
can be abated by building a zinc tube of some three inches
diameter into the chimney-stack, alongside the hot flue, or
inserting an iron tube within the old flue, and leading its

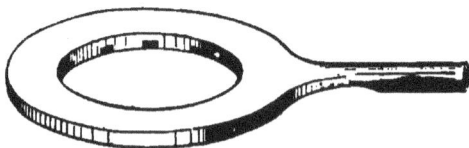

FIG. 7.—Annular ventilating-tube over bowl.

lower end into the space under the water-closet seat. In
wooden houses and other houses also, if in the country in
isolated positions, windy weather often crowds air up through
openings in the floor from the communicating floor-spaces, so
that in order to insure that the draught of this tube may draw
directly from the bowl of the water-closet under such circumstances, the tube should end in an annular flat tube of galvan-
ized iron, to be placed directly over the top edge of the bowl,
and under the seat, with perforations around the inner edge
of the ring, for withdrawing the air which we wish to get rid of.
These annular tubes are now made in Charlestown, and are
sold and applied by most plumbers. (See figure 7.) The
cover of the seat must then be arranged so that the valve can
be drawn after closing such cover, and care should always be
taken to so close the cover before lifting the valve. The foul
air which is puffed upwards at the instant of emptying the pan
is then sucked up by the draught of the chimney, without

an opportunity of mixing with the air of the room. Where no warm chimney can be had near enough to be thus used, the draught-tube can be run directly through the roof, with some ventilating attachment at its top to encourage the upward draught of air. This will often work well in winter, when the air of the house is artificially heated, and tends to escape by its own buoyancy. But in warm weather, it is not so likely to be of use. At that time the chief reliance is upon open win- *Windows needed in summer.* dows and a free current of air through the house. For this purpose it is always advisable to provide at least a part of a window directly over every water-closet. The pan and hopper closets which are often found tucked into corners, under stairways, and in other dark places, without special ventilation into chimneys, are sure to become nuisances, and poison all their surroundings. It has been suggested that the iron re- *Ventilation for iron receivers not efficient.* ceiver of the pan-closet should have a vent-tube between the pan and the lower trap, so as to provide for the exit of the foul air displaced by the descending charge of water. But this displacement is so sudden, and in such a large volume, compared with the capacity of such tube to receive it, that this does not appear feasible. Moreover, if the inner surface of the receiver is tapped by any sort of tube, its orifice would soon be likely to become smeared and stopped by the wet paper and fecal matter which is dashed about.

If an upward air-draught can be secured from the bowl, it is constantly at work, removing not only the foul air discharged when the pan is tipped, but all exhalations from water standing in the pan. This water is always exposed to the *Gases escape through water by solution and diffusion.* foul air of the receiver, around the outside of the bowl, and dissolves a certain amount of the gases from this source, to be given off and scattered by rapid diffusion in the air above, even when at rest, for the air is as a vacuum to other gaseous bodies. No vent from the iron receiver could remedy this, for that is foul past redemption. No circulation provided by such a vent could much affect the degree of its foulness.

The above-described defects in the pan-closet may in some degree be remedied by an efficient air-draught, in cases where this closet is already in use, but a surer remedy yet is found

D

in another style of closet, which can be applied with little extra cost when building anew. The closet made by George Jennings of London (see figure in margin) has accomplished the much-desired end of dispensing with the pan entirely, together with the air space between the bowl and the lower trap. It also dispenses with a separate trap below, having such a trap in itself, made in connection with the bowl, all in one piece of crockery. Baldwin Latham calls it "a perfect sanitary appliance." Its water-supply is taken directly from any supply-pipe, adjustable to the actual pressure, so that no separate tank, or service box or valve, wires or cranks are needed. The ordinary pan-closet alone costs less than half as much as Jennings', but its cost, with all those accessories, set up in working order, would be nearly as great. A supply direct from the pipes in the ordinary pan-closet is objected to with reason from the risk of back-flow of foul air from the closet into the pipes in case of lack of water-pressure from any cause at the moment of using the closet. But in the Jennings closet this risk is entirely avoided by the construction of the valve, which is a flap-valve, made of a rubber disc, rendering all back-flow impossible, and opening only with the pressure of water. Ample flushing of the bowl is secured by having the valve worked by a float, so that it remains open till the water reaches the prescribed level in the bowl. The Jennings closet does not seem to be quite all that could be desired, but it is certainly the best thing in the market. Its weak points are,—

First. The hollow plug, made hollow to act as an overflow for possible surplus of water delivered, allows the free escape of noxious gas, if any such there be, from the contents of the trap below. The only protection against this, as the apparatus is now constructed, would be the second lifting of the handle

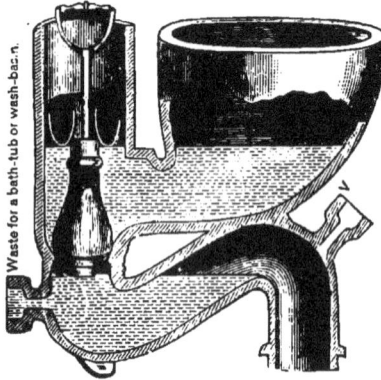

Jennings Water-Closet.

every time the closet is used to insure the complete expulsion of the foul matter from the lower trap.*

Second. Most of these closets in our market have no provision for a vent-hole in the trap. The necessity for such a vent was pointed out by Mr. Rogers Field, C. E.,† and a vent-hole is now provided, when demanded, by the makers at the point marked V on the diagram.

The cut should show the plug to be hollow, to act as an overflow. The draughtsman drew the plug in elevation, while he should have drawn it in section.

With this closet, the use of a disinfecting fluid, or the special ventilation of the closet-seat will probably be needless. But it does not escape the need of giving a vent to its own trap, as above described.

A copious vent for the soil-pipe itself should never be omitted.‡ Vents for soil-pipes.

The reason for this vent is as follows :—

The main drain of the house is supposed to be provided with a large trap, outside the house, as described above. There is also to be a trap under the water-closet, or forming a part of it. Between these two traps there must always be a confined column of foul air ; this column, if not connected with the outer air by a vent made for the purpose, is subject to compression or tension from the following causes, acting together or separately. Compression is caused,— Why always needed.

* Since the above was written, the writer learns that this defect has been entirely remedied by Mr. Jennings. He now attaches an inverted cup to the handle, just above the hollow plug, which effectually traps this air-hole. Parties ordering Jennings closets, should see to it that this important improvement is not omitted.

† The following is taken from a private letter of Mr. Field to the Secretary of the State Board of Health, in speaking of such a vent-pipe : "The function it has to perform is simply that of admitting air whenever the closet is worked so as to prevent the water being sucked out of the trap by the partial vacuum that would otherwise be created by the sudden rush of water down the arm leading from the closet to the soil-pipe. This pipe in no way does away with the necessity of having the soil-pipe properly ventilated ; and, *vice versa,* the ventilation of the soil-pipe by carrying it up above the roof does not do away with the necessity of this air-pipe."

‡ Mr. Simon, in the report above quoted, says : "That every private drain having inlets within a house, must have ascending from its head or heads into some suitable high position in the open air, and where it cannot infect the interior, a ventilating pipe or ventilating pipes of sectional area amply proportionate to its own."

First. From changes of temperature, either from change in that of the surrounding air, or by the pouring of hot water into the pipe.

Second. From the blowing of air into the soil-pipe from the sewers, which may not always be ventilated, and which may occasionally find the disconnecting trap disabled from some accident.

Third. From the influx of a considerable volume of water into the column from above, forcibly displacing an equivalent volume of air.

When compressed from either of the above causes, the foul air is blown out into the house at the orifice of some waste-pipe in connection, in spite of its trap.

Tension may occur from the reduction of temperature, or from the efflux of water as it leaves the confined column through the outside trap. In either case the vacuum is supplied by sucking the water out of some of the connecting traps, leaving their waste-pipes unsealed. The remedy is simple enough, and is often applied now in new buildings by carrying the soil-pipe up through the roof, with an open end, to connect the interior with the open air. In small houses, having but one or two water-closets, it will answer the purpose to apply a two-inch lead or iron pipe, to run from the top of the trap of the upper closet, up through the roof. If, however, the outside disconnecting trap is not ventilated, as above recommended, and if such small houses empty their drain into a public sewer, *nothing less than the whole size of the soil-pipe will be safe for its vent.* In building new houses, it is recommended that the soil-pipe be carried beyond the roof, and of its full size. If the sewers are ever tide-locked, a heavy rain at such times displaces an immense volume of air, which is forced into the house-drains, causing these vents to be fully taxed.

Where several water-closets are placed one above another, on different stories, drained by a perpendicular soil-pipe, as often occurs, it is not enough to extend the soil-pipe up through the roof. The trap on each of the closets below the upper one, except perhaps the very lowest, if this be at the bottom of the column, must have its own separate vent, otherwise the rush of water down the column from the upper

[margin note: Soil-pipe to be carried up through roof.]

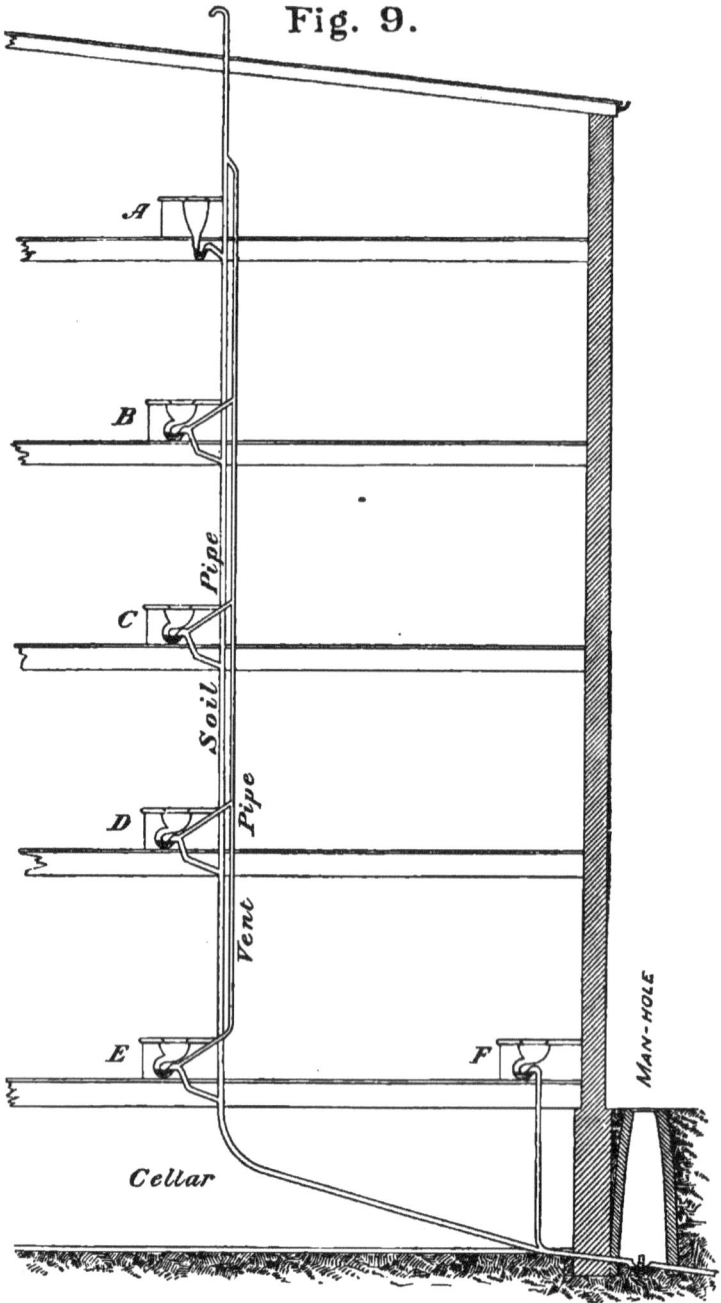

Fig. 9.

A

B

C

Soil Pipe.

D

Vent Pipe

E

F

Cellar

MAN-HOLE

story, or from any of the closets above the lower one, will be likely to siphon the water out of the intermediate traps in passing, or the trap itself which is used. The vents for this purpose should be at least two inches in diameter, and may all branch into each other and into the soil-pipe above the upper closet, as shown in the annexed diagram, Fig. 9. Thus, if we suppose a pailful of water to be emptied into the slop-sink on the upper floor at A, its rapid fall through the vertical soil-pipe would be likely to take with it by friction the air in the branches draining the closets at B, C, D and E. The traps of these closets would be likely to be drained by this siphon action, as above described, or by their own use, unless provided with vent-pipes, as shown in the figure. A closet placed as at F, connecting with the main soil-pipe where the latter is not vertical, is not subject to such action from use of those above, and has been found by actual experiment to be unaffected by their use when the main soil-pipe has an open top, as here shown. It would therefore probably not be necessary to apply a special vent-tube to its trap unless the vertical pipe immediately below it is several feet in length. If such be the case, the discharge of this closet itself might siphon the water out of the trap behind it, unless such a vent be provided, so that it is hardly safe to omit the vent in any case.

For those places where the cost of a Jennings water-closet is felt to be a burden, a simple hopper-closet is the best substitute. Simpler water-closets. But it should be properly ventilated under the seat, and its trap ventilated, and good provision made for flushing. This closet is described in Fig. 8 (p. 454), and consists of a hopper or bowl of crockery, set over a lead trap, or, what is better, a bowl with crockery trap in one piece. This closet gets rid of the confined chamber of foul air which condemns the pan-closet, and this is also easily cleaned, and simple. Its only fault is that the contents of the trap are directly exposed, so that trouble would ensue if ample flushing were not provided. It is sometimes provided with a constant flow of a driblet of water, which is both wasteful and inefficient. The flushing-water is needed only at the time of using the closet, or rather when leaving it, and it should then be applied in a sufficient quantity to drive the contents of the trap entirely through into the drain below. If a tank or service-box be applied to flush it, a definite supply of water can always be insured without such a waste as would render its use objectionable. In the way it is now used, the large amount of water wasted has led to the imposition of a special tax upon hopper-closets by the Boston Water Board. This waste, however, is an abuse, and not a necessary contingent upon the hopper-closet. The aperture at the bottom should be limited to about three inches in diameter, to prevent the admission of substances which might choke the drain.*

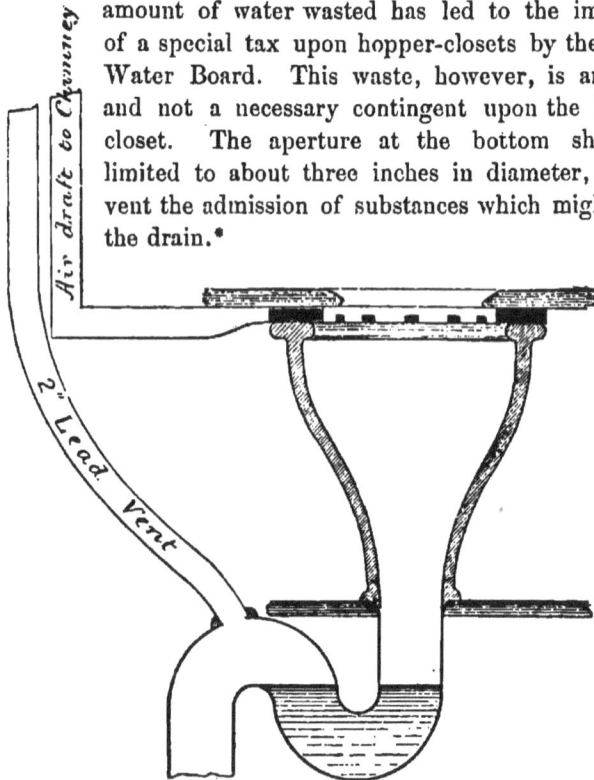

FIG. 8.—Hopper-Closet.

Direct supply of water-closets.

It has been usual with many plumbers of late to recommend the flushing of water-closets of the common "pan" or "hopper" style by a valve to be opened in a branch of

---

* In order to insure a sudden delivery of water to expel the foul contents from the trap of a hopper-closet, the service-box should have a capacity of at least a gallon, with a funnel-shaped bottom. The valve should be not less than two inches or two and a half inches in diameter, and the pipe leading the flushing-water to the hopper should be two to two and a half inches diameter, according to its length. Without such special provision for the sudden dash of water in a considerable volume, the exposure of foul matter in the trap would be sure to give trouble, as is the fact in all those in common use.

the main water-supply of the house, from which drinking-water is drawn through another faucet. But the only safe way is to break this connection by providing a small tank and service-box for the closet. With the valve furnishing water direct from the main, the following risk is incurred : The water-supply, even though nominally "constant," is some-times shut off temporarily for repairs in the street, and if at such time the water-closet valve should be opened, the air is drawn rapidly into the water-pipes from the interior of the closet, which is filled with organic vapors, and perhaps with actual contagium from disease. The water is soon let into the pipes again, mixing with this air and dissolving a portion, thereby becoming contaminated and unfit for drinking. This direct connection-valve has become so popular that one of the prominent dealers in plumbers' supplies recently alluded to the service-box supply as an "old-fashioned way which was going out of use," while conversing with the writer. In the supplementary report of the medical officer of the Privy Council for 1874, there is an interesting report from Dr. Buchanan, upon an outbreak of enteric fever in Caius College, Cambridge, where fifteen students were attacked while living in a building which was supposed to be provided with the most perfectly arranged sanitary appliances. After a most painstaking investigation, the fever was traced, by convincing evidence, to the use of a water-closet with direct supply from the mains, which had poisoned the water used for drinking in precisely the manner indicated above. In the same report, the following extract is found from the regulations under the metropolis water act of 1871 : "These regulations having for their object the pre-venting of undue consumption or contamination of water, are the result of an inquiry made for the Board of Trade by Lord Methuen, Captain Tyler, and Mr. Rawlinson, C. B. They have the sanction of the Board of Trade, and may be put in force by the London water companies." "Every boiler, urinal and water-closet in which water supplied by the com-pany is used (other than water-closets in which hand-flushing is employed) *shall*, within three months after these regula-tions come into operation, *be served only through a cistern or service-box*, and without a stool-cock, *and there shall be no*

[marginal notes:] Fever at Caius Col-lege, Cam-bridge.

London reg-ulations for indirect sup-ply to water-closets.

direct communication from the pipes of the company *to any boiler*, urinal or water-closet." "No pipe by which water is supplied by the company to any water-closet shall communicate with any part of such water-closet, or with any apparatus connected therewith, except the service-cistern thereof."

PUBLIC PRIVIES.

Public privies needed. The need of a system of public privies for the crowded parts of large towns is a subject inviting the earnest attention of all. who are interested in sanitary reform. All decent people who have occasion to frequent the narrow and crowded alleys where the poorer part of the people are lodged, will often be reminded of the streets of Rome and other European cities by the want of decency there prevailing. Among the various appliances used in different towns for this purpose, all have failed, unless where kept under the supervision of the local authorities. One thing is certain: the removal of filth is imperative, and it cannot be left for the people who use such public privies to take care of the apparatus. Under the same class may be considered the privies of jails, asylums, and other public institutions where large numbers are housed whose habits of cleanliness cannot be relied upon. For this purpose the water-carriage system is quite as applicable as in private houses, if only suitable apparatus be provided, and if it Liverpool trough-closets. be supervised by local authorities. Such apparatus has already been perfected in Liverpool, and in Bristol, England. Similar apparatus has lately been introduced in the schools at Dantzig, where the climate is about as rigorous as in Massachusetts. Annexed to the report of the medical officer of the Privy Council for 1874 is an interesting report by Mr. J. Netten Radcliffe on the means used in various towns for removal of excrement. Among his "conclusions" is the following (p. 154): "As regards the parts of a town or village inhabited by the poorer classes, a water-closet system may be managed so as to be entirely applicable to the circumstances of the most ignorant and most careless population. Essential conditions of such applicability, however, are, that the structural arrangements should be adapted to their purpose, and that the management should be wholly undertaken and efficiently done by the servants of the sanitary authority.

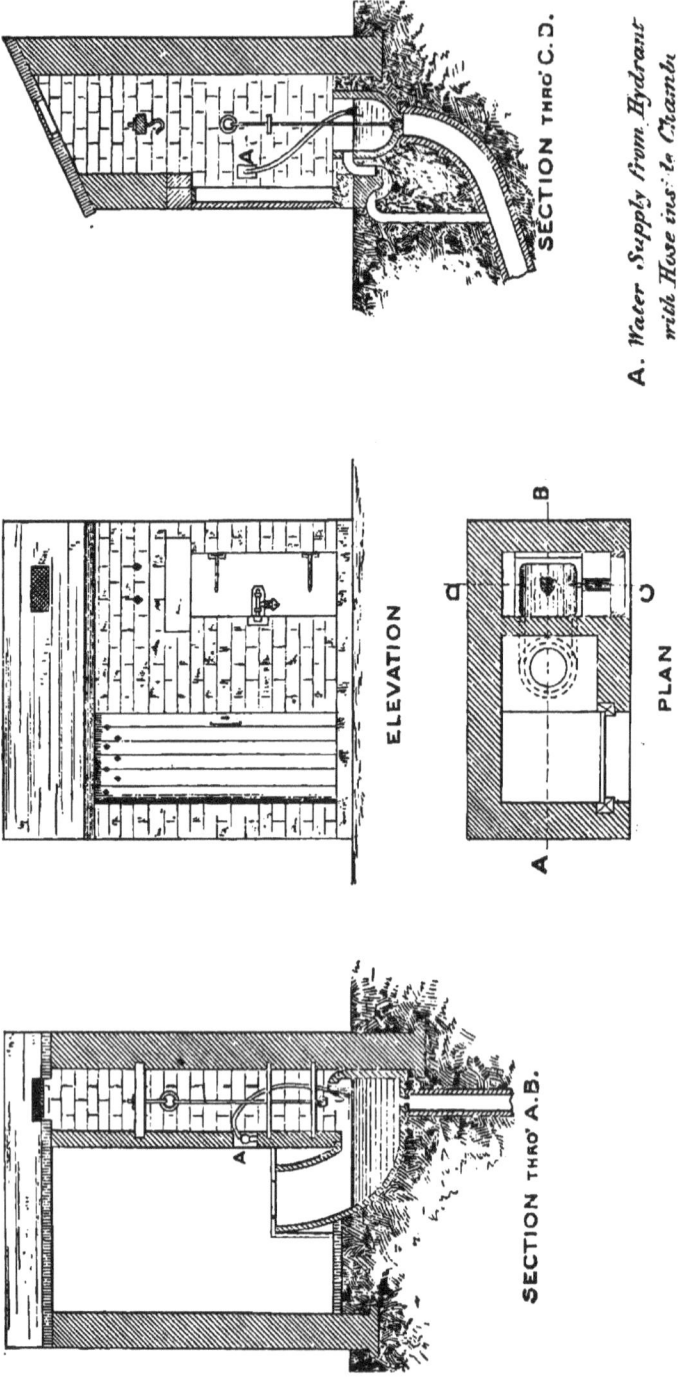

LIVERPOOL CORPORATION. TROUGH WATER CLOSET.

FIG. 10.

ELEVATION

PLAN

SECTION THRO' A.B.

SECTION THRO' C.D.

A. Water Supply from Hydrant with Hose insert to Chamber

LIVERPOOL CORPORATION. DOUBLE TROUGH WATER CLOSET.

FIG. 11.

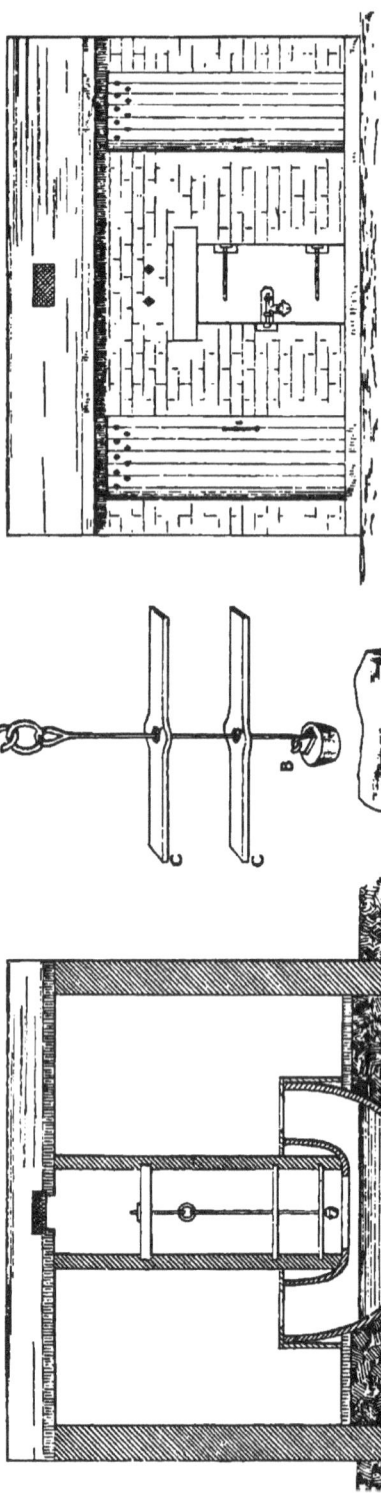

SECTION THRO' A.B.

ELEVATION

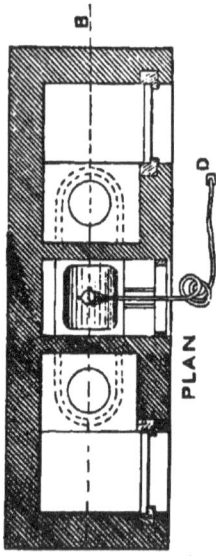

B. Enlarged Drawing of Valve,
   Guide rods (C) &c.

D. Water supply with Hose
   from Hydrant fixed in Court.

PLAN

Scale of Feet.

Where these conditions are observed as thoroughly as they are observed in parts of Liverpool and Bristol, water-closets are the best means of removing excremental matters from the poor neighborhoods of a town." In speaking of the introduction of these improvements in Liverpool, the same report says (p. 206) : "The council, in order to secure uniformity of action, and likewise to prevent future mistakes in the application of remedial measures, directed the town clerk to notify the several owners against whom proceedings should be taken, 'that it appears to the said council of the said borough, that the only effectual remedy for such privies and cesspools is by converting the same into water-closets.'"

"Since Dr. French has been medical officer, and mostly since 1866, he has ordered and obtained the conversion of 14,393 privies into water-closets ; and there were in 1869, in Liverpool, 20,000 privies attached to ash-pits and 31,150 water-closets, 2,150 of which are tank or trough closets." "These closets are constructed on a pattern ordered by the corporation, and approved, as to details, by the borough surveyor. Now, in 1874, the number of troughs for trough-closets is 3,304, serving for about 6,000 closets, and the number of water-closets other than trough-closets, 43,395." In order to explain more definitely the nature of the trough-closet, the annexed plates are copied from the same report of Mr. Radcliffe (figs. 10 and 11), with the following remarks (p. 206), viz. : "There is peculiar interest in the arrangement and working of the trough-closets which are in use by numerous families in the sort of neighborhoods where in other towns ordinary water-closets are commonly a failure and a nuisance. It remains to say that the position chosen for these new closets has been carefully determined by the circumstances of each place where they have been erected, and that peculiar facilities for their being well placed have been obtained by the time of their erection concurring with that of other improvements. The closets that are common to several families are cleansed in rotation by the people using them, and a register is kept of the order in which this should be done. Inspectors visiting the closets every two or three days see that this duty is performed, and are themselves held responsible for any shortcoming. By a little patience and firmness the inspector

B

succeeds in obtaining the necessary cleansing even among the most intractable classes, with very little assistance from the law. He will, if necessary, wait and see the closet cleaned out by the proper person. Last year only a dozen or so of people were summoned for neglect in this respect, and three of the offenders had to be sent to prison. It will be seen from the drawing that in connection with these closets there is au opening of access to the trough and water supply. This opening is for the scavenger, and the people using the closets have no concern with it. The scavengers are employed by the corporation, and every day they visit each of the trough-closets, unlock the iron door of access, discharge the contents of the trough, flush it out with hose and water, sweep it thoroughly clean, and leave it charged with fresh water for the next twenty-four hours' use. Frost has done no harm to these trough-closets, nor yet to the ordinary siphon-closet with its service-box." "There can be no question of the admirable efficiency of the working of the arrangements above described in the semi-public privies, nor of the recognition by the people of the superiority of the new to the old arrangements. Nor can there be any question that these results are due even more to the management of the whole business by the public authority than to the excellence of the constructive arrange-ments themselves. And not only is complete freedom from nuisance obtained where formerly filth and stink were uni-versal, but Dr. French states that in 1868, when an epidemic of enteric fever was prevailing in and about Liverpool, 'the only localities that seemed exempt from it were the places occupied by the poor, in which we had removed all the privies and made trough water-closets.'"

Traps. The use of traps on every waste-pipe inside of a house is a point upon which some difference of opinion is found. If the outside trap upon the main drain is well constructed and ven-tilated, there is somewhat less importance to be attached to those on each separate waste-pipe. Dr. O. Reynolds, in his little work entitled "Sewer-Gas, and How to Keep it out of Houses," thinks they may in most cases be safely dispensed with. But every waste-pipe becomes lined with a slimy film which, in a climate subject to such summer heats as ours, must needs decompose and give off offensive effluvia. The difference of temperature in different parts of a house, and the *Traps should be used at every waste inside a house.* winds outside, would always keep up drafts through waste-pipes if not trapped, passing down one orifice and up another, so that a dwelling-house can hardly be considered safe with us unless traps are provided at every waste somewhere within ten feet of its orifice. The chief objection to traps, except their cost, is that they delay or hinder, to a certain extent, the rapid efflux of the sewage, and keep a small quantity of it shut up to decompose within themselves. Moreover, there is always a column of confined air in a waste-pipe between any two traps in the same line of drainage. Care must therefore be taken to give this column a free connection or vent to the atmosphere, so that its tension may always be in equilibrium with the atmospheric pressure. Without this system of vents the traps are worse than useless, and deserve their name from the disappointment that would be sure to follow their use.

The best form of trap is that which gives least obstruction *Form of traps.* to the flow of the sewage, and requires the least quantity of water to insure its seal. All square corners tend to promote an accumulation of sediment or slime, which should be avoided. No form of trap is so simple and so generally efficient as the ordinary S-bend. Its calibre should be about as large as that of the pipe of which it forms a part, with continuous, smooth lines. When a waste-pipe is expected to carry at times a full stream of water, as from slop-sinks, wash-trays, and bowls, there is risk of having the traps siphoned by the last water passing down, unless they be provided with a vent-hole at the top, and this must of course have a pipe to carry its effluvia to the top of the roof.

The position of traps is a matter of some importance. If *Position of traps.* close to the orifice, as in the common sink bell-trap, a slight obstruction of sediment will soon serve as a nucleus for a complete dam; but if two or three feet below the orifice, and directly under it, such a slight obstruction would soon be swept away by accumulation of the two or three feet head of water above. The common bell-trap used in sinks, and at-*Bell-traps of no use.* tached to the strainer, conforms to neither of the above conditions, and is a mere subterfuge. Being close under the strainer, no head of water can accumulate to flush it, and its large annular depression is a receptacle of rubbish, In fact, it is a mere obstacle to the drainage, so that most cooks lift the cover when they want the water to run off, losing the benefit of both strainer and trap together. A strainer is useful, if screwed fast down, with holes of ample size and number, and a trap made of the S form about the level of the floor, under a wash-tray or sink, will run for years and keep itself clear of sediment, with such a strainer over it.

WORKMANSHIP.

Workman-
ship.

This subject has been mentioned above, but its importance may justify something further. There is generally but one way in which to do a thing properly and well, while there is an endless number of wrong ways into which workmen stumble, through ignorance or recklessness. The importance of having the best quality of work in matters that so nearly affect the health and lives of our families, need not further be dwelt upon. These two facts confront us. Good work is the Alpha and the Omega of good drainage, and good work is too rare an article among the workmen in mechanical trades. Two hundred years ago, the artisans' guilds of the north of Europe developed a degree of skill which often made an artist of the artisan, producing such men as Albrecht Dürer and Peter Fischer, of Nürnberg, while the general emulation led to a degree of excellence in workmanship among large numbers that calls for our admiration, and gave a tone to the

Too little
emulation
for excel-
lence.

civilization of that age well worthy of imitation. Our modern "trades unions," however, seem to have encouraged a lack of thoroughness and lack of interest in the artisan by making too little distinction between the faithful and the unfaithful workman. They combine for mutual protection against evils which they themselves tend to aggravate, and are led by demagogues to fix arbitrary rates of wages, regardless of merit, and encourage no emulation for excellence. The consequences are bad enough, morally, in the lowering of the standard of excellence among artisans, and thereby degrading their intellects, as well as their morals; but the results are particularly disastrous among those who employ artisans to construct such important works as house-drains, for the average householder necessarily depends largely upon the intelligence and skill of the artisan for the perfection of details in such work, concerning which he, the employer, is ignorant. Of course the blundering and reckless apprentice becomes, in time, the ignorant and stupid master-workman, deficient in skill, and confined to routine.

Instances often come to our notice like the following : Some years since the writer employed a man sent from one of the best plumbing establishments in Boston to attach a vent-pipe to a soil-pipe in an old house, which had been built, like thousands of others, without one. The workman was directed to lead the vent into the kitchen chimney-flue, which was conveniently near. This pipe was afterwards seen, entering the side of the chimney-breast, and appeared all right. A few weeks since, a new tenant in the house complained of a bad smell. After searching all about for its source, suspicion led to an examination of the connection of this vent with the chimney. It was found that it *never entered the chimney at all*, but ended with an open mouth in the furring-space between the chimney and plastering ! Of course all the gases passing out of this vent, had for years had free access to all the floors of the house. Luckily nobody had had typhoid fever ; but no thanks are due to the plumbing for the immunity.

Cases of unfaithful work.

Another instance of bad faith in a laborer occurred within a few months in the same street. A recent introduction of aqueduct-water led to laying a service-pipe into a house where it happened to cross under the drain in the front yard. The foreman was cautioned about the drain, which was of Scotch pipe. Some months afterward the occupants of the house found filth oozing through the front cellar wall of the house in midwinter. On digging down through four feet of frozen ground in the front yard, the Scotch drain-pipe was found to have been broken open and placed together again without using any cement, while putting the water-pipe through under it. No mention being made of this fact, it was not suspected till the leakage made its way through the cellar-wall, as above related. Men who are guilty of such acts of bad faith are not likely to be employed further by the same parties ; but some more serious penalty than loss of patronage seems to be needed for such cases, where the results may be the illness and death of their fellow-citizens and neighbors.

Duties of architects. The want of skill among artisans has been aggravated by a degree of ignorance on the part of architects upon points where they are expected by the community to be experts. It may be urged in their defence that house-drainage is a comparatively new luxury. If so, the architects should be the very class who, from their position, should do the most towards developing its perfection, and not leave it, as is often done, for the confused notions of their clients to work out the details, with the advice of the head mason, neither of whom have probably ever studied the subject from any point of view more comprehensive than from their own limited experience. Much has been written on these subjects; so much, that but few new ideas have been brought forward in this paper. Those who wish to study it further will find it more amply treated by such men as Bazalgette, Latham, Corfield, Eassie, Menzies, Parkes, Reynolds, Waring, Shedd, and others, all in our own language, besides as many more in German and French, wherein the peculiar stand-points of the various writers give rise to different treatment, each suited to its own locality or climate. Besides the study of the experience of others, much remains to be done to adapt means to ends, taking into account the peculiar circumstances governing each case under treatment. What the community has a right to expect, is, that men having in hand the designing and erection of their dwellings, should inform themselves of the conditions on which such dwellings can be made healthy places of abode. If they did this, as a rule, and if artisans had, as a rule, that pride of character and love of good workmanship which is the capital of the mechanic, we should soon see better results.

The writer does not intend to ignore the fact that there are many intelligent and painstaking architects, who have given this subject careful attention; but for some reason or other, there are many of our important buildings designed and erected under the charge of those who are not so distinguished. The following examples came under the writer's notice, within a short time, which may serve to illustrate this point.

A conspicuous public building, costing nearly $200,000, Mistakes in planning. was lately erected under the charge of a leading architect. Like most similar buildings of the present day, it contains a good deal of plumbing. For some time past a nauseous odor pervaded the cellar, which no amount of window opening would remedy. It increased to such an extent as to fill the whole building, and render its occupation sickening. On inquiry, it was found that some square cesspools had been constructed under the cellar floor, into which various drain-pipes entered, and from which a brick drain led, covered with flat stones, an invention of a past age. One of these reservoirs of filth under the cellar floor had been forced to overflow, and had saturated the concrete pavement, after a temporary obstruction of the outlet by a careless workman. Although the obstruction was promptly removed, the stench remained a long time. What good purpose the cesspool served, or is capable of serving, is past comprehension. Its presence is a mere nuisance. The brick drain could hardly have been planned by a person who had taken pains to learn the inherent faults of such structures, and the great superiority of smooth pipes.

Another case : a large public building was recently planned by an architect, chosen by the parties who had the subject in charge, as peculiarly versed in the wants of such an institution as this was to accommodate. The plans showed a brick drain, big enough to crawl through, running under the building for a length of over two hundred feet, too large to be self-cleansing, and, therefore, merely a prolonged cesspool, over which a population of about five hundred persons were to be lodged, those living on the lower story being on a stone floor, resting on the ground, without a cellar or any subsoil drainage, except the brick sewer! This was not all. A large tank or cesspool was provided, just outside the building, in which the whole sewage of this population was delayed, to settle and ferment till convenient times for its removal, although a good sewer was at hand for its immediate and rapid transit to a safe distance.

DUTIES OF PROPRIETORS AND OCCUPANTS.

Duties of proprietors. However well a system of house-drainage may be planned and constructed, it cannot be expected to be entirely automatic, or to serve its owner for an unlimited period without *Care required.* intelligent supervision. In fact, "eternal vigilance" is the price of safety in such matters in a climate where such violent and sudden changes occur as in ours. Sometimes a trap *Dangers from frost.* may freeze in January and dry up in July. Deep frosts sometimes break up drains, and leave them leaky. Rats burrow into and gnaw into drains, if not thoroughly built. The gases given off by sewage often corrode lead pipes, and the ammonia in water-closets corrodes the copper pans. Valves become leaky by wear. Counterpoises get loose. But frost is our greatest enemy; a frozen water-pipe often does much damage, but a frozen drain is the climax of discomfort. With the extended use of plumbing come the increased risks of such mishaps, till many householders long to simplify the *Record of drains needed.* apparatus. It cannot certainly be well taken care of in country houses in our climate, unless the occupant knows where to find the pipes, and how to empty them on frosty nights. The risks of leakage of drains are of course very serious, and the difficulty of tracing such troubles to their sources renders it imperative to keep a careful record of their position, and to take the alarm from the only sense by which we can often be led to detect them, acting vigorously to repair the defect when found. Those who do not wish to trouble themselves with such matters had better dispense with drains entirely, and do as in the days of our fathers; viz., carry the refuse-water to a safe distance from the house in pails, where it can be consigned to mother earth. Then they can feel sure that it is beyond the chance of harming them in the house.

www.ingramcontent.com/pod-product-compliance
Lightning Source LLC
Chambersburg PA
CBHW022025190326
41519CB00010B/1607